YOUR KNOWLEDGE HAS VALUE

- We will publish your bachelor's and master's thesis, essays and papers

- Your own eBook and book - sold worldwide in all relevant shops

- Earn money with each sale

Upload your text at www.GRIN.com and publish for free

Bibliographic information published by the German National Library:

The German National Library lists this publication in the National Bibliography;
detailed bibliographic data are available on the Internet at http://dnb.dnb.de .

Imprint:

Copyright © 2018 GRIN Verlag
Print and binding: Books on Demand GmbH, Norderstedt Germany
ISBN: 9783668703278

This book at GRIN:

https://www.grin.com/document/425391

Joshua Lawrence Langat

The Historical Conflict and Implication of Evolution and the Science on Contemporary Education

GRIN Verlag

THE HISTORICAL CONFLICT AND IMPLICATION OF EVOLUTION AND THE SCIENCE ON CONTEMPORARY EDUCATION

BY

Dr. Joshua Lawrence Langat

May, 2018

Contents

1.0 Introduction

This chapter examines the background information to the study, the evolution of man - scientific evidence, the scientific reception of Darwinism (Darwin's Theory of Evolution - the premise Darwin's theory of evolution - natural selection Darwin's theory of evolution - slowly but surely, Darwin's theory of evolution - a theory in crisis), metaphysical concerns on theory of evolution, methodological objections of theory of evolution, reconsidering the nature of science from physics to evolutionary biology, from empiricism, toward a naturalistic model of scientific practice and conclusion of the study.

1.1 Background of the study

The modern theory concerning the evolution of man proposes that humans and apes derive from an apelike ancestor that lived on earth a few million years ago. The theory states that man, through a combination of environmental and genetic factors, emerged as a species to produce the variety of ethnicities seen today, while modern apes evolved on a separate evolutionary pathway. Perhaps the most famous proponent of evolutionary theory is Charles Darwin (1809-82) who authored The Origin of Species (1859) to describe his theory of evolution. It was based largely on observations which he made during his 5-year voyage around the world aboard the HMS Beagle (1831-36). Since then, mankind's origin has generally been explained from an evolutionary perspective. Moreover, the theory of man's evolution has been and continues to be modified as new findings are discovered, revisions to the theory are adopted, and earlier concepts proven incorrect are discarded.

Evolution of Man - Scientific Evidence -The theory of evolution of man is supported by a set of independent observations within the fields of anthropology, paleontology, and molecular biology. Collectively, they depict life branching out from a common ancestor through gradual genetic changes over millions of years, commonly known as the "tree of life." Although accepted in mainstream science as altogether factual and experimentally proven, a closer examination of the evidences reveal some inaccuracies and reasonable alternative explanations. This causes a growing number of scientists to dissent from the Darwinian theory of evolution for its inability to satisfactorily explain the origin of man.

One of the major evidences for the evolution of man is homology, that is, the similarity of either anatomical or genetic features between species. For instance, the resemblance in the skeleton structure of apes and humans has been correlated to the homologous genetic sequences within each species as strong evidence for common ancestry. This argument contains the major assumption that similarity equals relatedness. In other words, the more alike two species appear; the more closely they are related to one another. This is known to be a poor assumption. Two species can have homologous anatomy even though they are not related in any way. This is called "convergence" in evolutionary terms. It is now known that homologous features can be generated from entirely different gene segments within different unrelated species. The reality of convergence implies that anatomical features arise because of the need for specific functionality, which is a serious blow to the concept of homology and ancestry.

The importance of evolution to the biological sciences is a point that has been acknowledged in the science education literature. Indeed, the standard practice has been to pay obeisance to its centrality by citing Dobzhansky's (1973) well-known statement regarding the sense making powers of Charles Darwin's theory in the opening paragraph of articles on the subject. The enthusiasm, with which this point is embraced, however, has not been matched by a corresponding increase in scholarly attention. Evolution remains a relatively under researched topic within the science education community (Cummins, Demastes, & Hafner, 1994). The lamentable fact remains that "large numbers of people reject the theory of evolution, and the science education community has done little to help teachers present evolution in a way that will ameliorate this situation" (Smith, Siegel, & McInerney, 1995, p. 23). This state of affairs presents problems not only for the teaching and learning of evolution, but also for the public perception of science as a whole, of which evolution is clearly the most publicly suspect representative.

Although the field has recently shown signs of life, one of the reasons for its neglect, apart from its factious history in the United States, may stem from the seemingly difficult and amorphous nature of the subject matter itself. It is fairly well established that most students, as well as members of the general public, view the biological world from a kind of pre-Darwinian perspective.

This has been documented by researchers in science education and has commonly been taken as the starting point for developing new strategies for teaching evolution in schools (Demastes, Trowbridge, & Cummins, 1992). One approach that has demonstrated promise involves introducing students to the historical precursors of Darwin's theory. The success of this technique, indeed for a good number of the instructional interventions tried so far, has nonetheless been limited (see, for example, Jensen & Finley, 1996). Broad-scale student mastery of evolutionary theory has proven to be quite elusive—a conclusion uniformly reached in the science education literature (Demastes et al., 1992). It has also been reported that even a significant number of science teachers have serious questions when it comes to evolution (Eve & Dunn, 1990).

The pervasiveness and persistence of these difficulties suggest, perhaps, that we look for a deeper explanation than any so far considered. The historical parallels between student experiences in the biology classroom and the experiences of 19th-century scientists are particularly instructive in this regard, though not for the reasons commonly supposed. There existed in Darwin's time a fundamental mismatch between the view of science endorsed by Darwin's contemporaries and the implicit model of science on which he based his theory of evolution by natural selection. The significance here lies not in the similarity between any initial student conceptions and early evolutionary ideas, but rather in the historical longevity of the mismatch between these two views of science.

Past characterizations of science, historically derived from physics, internalized a broadly empirical and experimental bias that failed to accommodate key issues evolutionary biology introduced to the scientific community. There has been little done over the past century to reconcile these views, especially in science education, resulting in a situation that continues to impede the development of effective instruction in evolution. The specific difficulties students have understanding evolution are less perplexing when considered in light of the resistance Darwin encountered from the scientific community of his own time. A careful examination of the scientific reception of evolution in the 19th century reveals interesting parallels with current accounts of student learning that suggest positive directions for new work in evolution education research.

1.2 The Scientific Reception of Darwinism

Darwin's Theory of Evolution is the widely held notion that all life is related and has descended from a common ancestor: the birds and the bananas, the fishes and the flowers -- all related. Darwin's general theory presumes the development of life from non-life and stresses a purely naturalistic (undirected) "descent with modification". That is, complex creatures evolve from more simplistic ancestors naturally over time. In a nutshell, as random genetic mutations occur within an organism's genetic code, the beneficial mutations are preserved because they aid survival -- a process known as "natural selection." These beneficial mutations are passed on to the next generation. Over time, beneficial mutations accumulate and the result is an entirely different organism (not just a variation of the original, but an entirely different creature).

Darwin's Theory of Evolution: Natural Selection - While Darwin's Theory of Evolution is a relatively young archetype, the evolutionary worldview itself is as old as antiquity. Ancient Greek philosophers such as Anaximander postulated the development of life from non-life and the evolutionary descent of man from animal. Charles Darwin simply brought something new to the old philosophy -- a plausible mechanism called "natural selection." Natural selection acts to preserve and accumulate minor advantageous genetic mutations. Suppose a member of a species developed a functional advantage (it grew wings and learned to fly). Its offspring would inherit that advantage and pass it on to their offspring. The inferior (disadvantaged) members of the same species would gradually die out, leaving only the superior (advantaged) members of the species. Natural selection is the preservation of a functional advantage that enables a species to compete better in the wild. Natural selection is the naturalistic equivalent to domestic breeding. Over the centuries, human breeders have produced dramatic changes in domestic animal populations by selecting individuals to breed. Breeders eliminate undesirable traits gradually over time. Similarly, natural selection eliminates inferior species gradually over time.

Darwin's Theory of Evolution: Slowly But Surely- Darwin's Theory of Evolution is a slow gradual process. Darwin wrote, "...Natural selection acts only by taking advantage of slight successive variations; she can never take a great and sudden leap, but must advance by short and sure, though slow steps" (Darwin, 1859). Thus, Darwin conceded that, "If it could be demonstrated that any complex organ existed, which could not possibly have been formed by numerous, successive, slight modifications, my theory would absolutely break down" (Ibid. p. 158).

Such a complex organ would be known as an "irreducibly complex system". An irreducibly complex system is one composed of multiple parts, all of which are necessary for the system to function. If even one part is missing, the entire system will fail to function. Every individual part is integral (Behe, 1996). Thus, such a system could not have evolved slowly, piece by piece. The common mousetrap is an everyday non-biological example of irreducible complexity. It is composed of five basic parts: a catch (to hold the bait), a powerful spring, a thin rod called "the hammer," a holding bar to secure the hammer in place, and a platform to mount the trap. If any one of these parts is missing, the mechanism will not work. Each individual part is integral. The mousetrap is irreducibly complex (Illustra Media, 2002).

Darwin's Theory of Evolution in crisis - Darwin's Theory of Evolution is a theory in crisis in light of the tremendous advances we've made in molecular biology, biochemistry and genetics over the past fifty years. We now know that there are in fact tens of thousands of irreducibly complex systems on the cellular level. Specified complexity pervades the microscopic biological world. Molecular biologist Michael Denton wrote, "Although the tiniest bacterial cells are incredibly small, weighing less than 10^{-12} grams, each is in effect a veritable micro-miniaturized factory containing thousands of exquisitely designed pieces of intricate molecular machinery, made up altogether of one hundred thousand million atoms, far more complicated than any machinery built by man and absolutely without parallel in the non-living world" (Denton, 1986).

And we don't need a microscope to observe irreducible complexity. The eye, the ear and the heart are all examples of irreducible complexity, though they were not recognized as such in Darwin's day. Nevertheless, Darwin confessed, "To suppose that the eye with all its inimitable contrivances for adjusting the focus to different distances, for admitting different amounts of light, and for the correction of spherical and chromatic aberration, could have been formed by natural selection seems, I freely confess, absurd in the highest degree" (Darwin, 1859).

There were a variety of responses to the 1859 publication of *On the Origin of Species*. These ranged from solid endorsement to arrant condemnation within the scientific community—to say nothing of the reaction of the general public. Among the critical responses documented, one can find two primary threads of objection. The first and most profound concerned the strong naturalism of Darwin's theory. The scientific community in Victorian England operated within a theistic context that presupposed notions of intelligent design and purpose in the world (Ruse, 1979). Darwin's theory of evolution by natural selection directly challenged the need for such metaphysical commitments and was understandably met with resistance for doing so. The second objection related to apparent methodological irregularities in Darwin's work that for many threw into doubt the validity of his conclusions. Both categories of objection stand out as significant not only historically but, more important for this article, for evolution education today. A brief summary of the historical reception will provide the background necessary for drawing out the implications for science education.

1.3 Metaphysical concerns on theory of evolution

The general scientific uproar surrounding the publication of Darwin's *Origin* had surprisingly little to do with the empirical claims on which his theory was based. It was uncommon to find disagreement over the basic assertions Darwin made about the variability of species, the survival rates of young, the plasticity of domestic animals, and the like. These claims and the abundant examples Darwin marshaled in their support were accepted without palpable controversy.

Rather, it was his philosophical approach as a whole and its consequences for the place of humans in nature that flew in the face of the prevailing assumptions of the time. Darwin's commitment to a completely naturalistic explanation of the origin and diversity of organisms, now seen as his most significant contribution to biology (Bowler, 1983), was then viewed by many as a dangerous materialism that eliminated divine agency from the history of life on earth. This notion of the divine reinforced deeper philosophical commitments to essentialism and teleology that presented formidable obstacles to the easy acceptance of evolution by natural selection.

Essentialism, the belief in the existence of discrete kinds each with a universal essence, lays claim to the deepest philosophical roots in Western thought (Hull, 1973). Early Aristotelians argued such kinds existed independently in nature, while Platonists believed such universal essences were located only in the realm of ideas. In one form or another conception of natural kinds, or types, was the foundation of early systems of biological classification in which species were identified and grouped according to their unchanging observable characteristics. Any deviations from the type were viewed as inconsequential developmental aberrations (Mayr, 1991). Such typological thinking was a key component in the anatomical theories of French and English biologists such as Cuvier and Owen in the first half of the 19th century and was also common to the German morphologists of the time (Nyhart, 1995; Russell, 1916).

1.4 Methodological objections on theory of evolution

Apart from the fundamental assumptions that made up the content of Darwin's theory was the methodology he employed in their justification—a point with which many of his contemporaries took issue. Although some methodological criticism derived, no doubt, from prior metaphysical objections to evolution, there were also scientists who, although sympathetic to Darwin's conclusions, were sincerely troubled by the means he used to arrive at them. The early 19th century was a time when many eminent natural philosophers in Britain were systematically laying down the foundations of scientific methodology (Hull, 1973). Classics in this field were John Herschel's *Preliminary Discourse on the Study of Natural Philosophy* (1830) and John Stuart Mill's *System of*

Logic (1843). These works set the standard against which Darwin's work was judged and, fairly or not, found wanting. The difficulties British scientists had reconciling the explanatory success of evolution by natural selection with its apparent methodological shortcomings reveal a great deal about the gap between the nature of science as practiced and the nature of science as perceived.

The problem, primarily due to an overly restrictive account of science, was compounded by an even less appropriate comparison across disciplines. Evolutionary biology in 1860 was thoroughly grounded in the long tradition of work in the field of natural history, the scope of which included rocks and minerals as well as animals and plants. Confronted with the enormous diversity of specimens in all areas, the field naturalist's first task was careful observation followed by description and systematic classification. Such work produced the great natural histories of the 18th century and continued as important scientific work into the 19th century as well. The essence of such work was descriptive and took nature as found. With descriptive work extending to the intricacies of embryological development and the succession of fossil life forms in rock strata, the importance of historical explanation as the organizing idea in biology took hold (Coleman, 1977). Naturalists looked to the past as the key to understanding the present, searching for patterns and laws of development. During this time, evolutionary ideas surfaced periodically in attempts to understand the history of life on earth, but such ideas had to wait for Darwin to become generally accepted by the scientific community.

1.5 Reconsidering the Nature of Science
Although the similarities between children's conceptions of natural phenomena and early scientific conceptions, in some cases, have been carefully documented and are at times striking (Brumby, 1984, for example), there is no reason to expect any necessary similarity between the subsequent path of conceptual development in the student and that in the history of biology. The highly contextual nature of historical and scientific change makes any such claims clearly unwarranted. What the history of biology can contribute to educational research, in this case, is a better understanding of the past intellectual conditions that impeded the scientific acceptance of Darwinism—conditions that, for

various reasons, have persisted over time and may influence student conceptions in ways that make understanding evolutionary theory difficult today.

Lessons we can take from the reception of Darwin's theory involve recognizing the difficulties that scientists had in making the transition to new conceptions of both the history of life on earth and the best means to be employed in its investigation. In the long view of history, these transitions are relatively recent and not entirely complete, especially outside the field of evolutionary biology; their remnants still pose problems for us even now. What is needed is to uncover the fundamental obstacles, both present and potential, to student understanding of an evolutionary picture of the world and examine current attempts at evolution education in their light. Those obstacles seem well represented in the historical survey above, which suggests that for students to appreciate the import and essence of Darwin's work, a reconceptualization of our view of science within the science education community is necessary. Such a view must include recognition of both the heterogeneous nature of disciplinary practice as well as the importance of nonempirical factors in scientific advancement.

1.6 From Physics to Evolutionary Biology

The movement in science studies toward viewing the various disciplines that comprise science as a set of distinct local practices rather than as parts of a unified system, though ongoing since the 1960s, has failed to make significant inroads into the work of the science education community (Galison & Stump, 1996). Most programs in education research have taken physics as their implicit model for all science without considering the inevitable distortions that result when that research framework is applied to disciplines outside physics. More problematic has been the tendency for science educators to use the physics model in the development of curriculum intended to address general issues concerning the nature of science, thus perpetuating a misunderstanding of scientific method that may foreclose the effective learning of evolutionary biology (Hodson, 1996; Jenkins, 1996).

The situation that exists in science education, not surprisingly, is a consequence of past work in the philosophy of science. Efforts to formalize scientific theories in the 20th century were shaped significantly by the work of the logical positivists (Joergensen, 1970). These philosophers, just as the early 19th-century methodologists before them, used the unquestioned success of physics as their paradigm for understanding all science (Laudan, 1984).

1.7 From Empiricism

Recognition of evolutionary biology as a science distinct from the generally narrow Physics-based model of science is a necessary but not sufficient first step toward a more robust treatment of evolution education and research. The disciplinary myopia described above has been accompanied by a limited philosophical view that has contributed to an overly rational, exclusively empirical view of science, which has handicapped both educational researchers and biology students with respect to evolution. Recent work in the field of science studies, with its emphasis on science as actually practiced, however, provides an initial framework that may be useful in developing a more productive research program in evolution education.

With respect to Darwin's theory, as we have seen, there was no easy resolution to the controversy that flared up in the aftermath of its presentation. Key points of debate centered on the adequacy of his method rather than any appeal to empirical evidence. Metaphysical commitments played a central role as well, as scientists either rejected the raw naturalism of the theory in favor of essentialist notions of creation or, more commonly, accepted the evidence of species transmutation while modifying the mechanism of natural selection to include elements of teleology in orthogenesis and Lamarckian use-inheritance. In either case, the historical record reveals the tenacity of these metaphysical ideas in the absence of any significant positive evidence. Issues such as these would be cast as irrational or nonscientific at least, if one were to accept the empiricist bias of the logical positivist program. However, to ignore this element of scientific debate is to impoverish and misrepresent the nature of theory choice in particular and the progress of science in general.

There has been some movement in the teaching and learning research toward a more inclusive account of the nature of science, especially in the area of conceptual change. Posner and Strike acknowledged past overemphasis on the rationalism of the model (1992) and Duschl and Gitomer (1991) argued for a more accurate model of scientific change for use in assessment, one that takes into account the aims and methods of science. Despite these theoretical modifications, the conceptual change model has proven to be somewhat problematic as a guide for instructional practice (Hewson & Thorley, 1989). This is true especially in scientific areas that fall outside the narrow conception of science on which conceptual change was originally based. Recent studies in the context of evolutionary biology—a subject with strong affective and metaphysical elements as we have seen—have demonstrated some of these limitations (Demastes, Good, & Peebles, 1995, 1996).

1.8 Toward a Naturalistic Model of Scientific Practice

In the same way that the growing incongruity between the history and philosophy of science in the 1950s sparked the post positivist shift begun by Kuhn, the inability of current research to address, let alone fruitfully engage, important conceptual difficulties students have with evolution suggests the need for the science education community to reassess its view of the nature of science. Rather than relying on the narrow empiricist, physics-based model of science that has limited our ability to deal effectively with evolution education, we need to examine broader models of science to help guide education research and curriculum development. What is needed is a view of science that not only represents all disciplines with their varied methodological approaches, but also will provide students with a more accurate view of the functional role metaphysical assumptions play in the advancement of science.

The most promising models can be found in the recent work of the science studies community. These have been developed from a close examination of scientific practice, either from firsthand observation or through the scrutiny of a more inclusive historical record (Siegel, 1993; see also Donovan, Laudan, & Laudan, 1988). Older philosophical models, as mentioned previously, sought to formalize the logical structure of science, to isolate its rational essence, in order to distinguish science from less objective fields of inquiry and so that its methods might be propagated more effectively. This attempt at decontextualization, it is now conceded, warped rather than purified. The result was often a scientific model that was, if logically consistent, at odds with reality (Laudan, 1984). Following what has been referred to as the naturalistic turn, philosophers and sociologists of science have recognized the importance of the broad socio-institutional matrix in which scientific practice takes place. Although no unified view of the nature of science has yet emerged, there is much to be found in the rich description of scientific practice that can be used in both evolution education and science education research (Jenkins, 1996).

In the field of science studies, one can trace the emergence of modeling as a central component of scientific practice. Giere described theory development in science as the development of a "population of models consisting of related families of models" that purport to represent the workings of the natural world (1988, p. 143). Drawing on work in cognitive science, he described scientific models as not much different from the mental models people use in their every- day affairs. Kitcher (1993) also saw the articulation of explanatory schemata, or models, as "a crucial part of a scientist's practice." The goal of such practice, he argued, is primarily to increase the adequacy of these models by improving the specificity of their key concepts as well as improving their overall organization. Pickering (1995), in a much more encompassing view of scientific practice, defined modeling as the sole means by which the current state of scientific knowledge is transformed into future states.

The work referred to above is by no means exhaustive, and there would likely be sharp differences of opinion regarding the details of scientific practice as understood by the authors cited here. However, we need not wait for the philosophical minutiae to be settled before we choose to make use of what has been generally agreed upon—especially as it represents a step forward from what has proven to be inadequate so far for the purpose of evolution education. All these authors describe a process in which some representation of the natural world is extended and refined in practice according to criteria determined by the scientific community. The representation can be a theory, which itself can be defined more precisely as a family of models, or an explanatory schemata. For our purposes the general term *model* will suffice and can be defined loosely as an abstract system that represents the workings of the natural world. Such a system is composed of related elements that interact in predefined and predictable, though not necessarily deterministic, ways. Left here—science as the modeling of empirical phenomena - this summary of practice would differ little from the traditional accounts that have been found wanting.

The naturalistic studies of science, however, reveal a much broader understanding of scientific practice, one in which scientific theories, or models, represent only one component of the more complex machinery of inquiry. In addition to theories, Kitcher included language, natural assumptions, methodological views, and canons of good observation and experiment in this multidimensional entity. In a longer view of scientific change over time, Laudan (1984) outlined a triadic network in which theories, methods, and cognitive aims interactively determine the course of scientific advance. None of these operates individually to advance science; theory choice is justified by methodological rules which are themselves subject to revision in light of the cognitive values or aims of the scientific community. These components, Laudan argued, are mutually interdependent; i.e., methodological rules can determine the range of possible theories, new theories can radically alter the nature of previously determined cognitive aims, and so on.

Over time, science progresses via a stepwise process in which two components of the triad always provide a stable context for change in the remaining component. Pickering similarly understood the production of knowledge to include theories, disciplinary structures, metaphysical assumptions, and the action of the natural world, adding the social organization of scientists as an important additional component. All of these, he stated, are elements of practice that are interactively stabilized, each contributing in varying degrees to the extension of scientific understanding.

Taken together, these works present a rich picture of the nature of science that is both more complex than past philosophical accounts of science and more authentic. They point to important aspects of practice that have been neglected in science education. In addition, they reveal that it is not enough merely to add to the mix discussions of scientific method, or metaphysics. Science education in various instances has seen calls for the inclusion of methodological discussions, the recognition of the social nature of scientific inquiry, and the acknowledgment of metaphysical points of view. Too often, though, these are viewed as extras to be included as time permits, or in courses designed for the less able to humanize and make the science more appealing. Teaching theories well, even with a detailed look at the empirical evidence that supports them, is not enough, in some cases, for a comprehensive understanding of those theories, or of science as a whole. The growing consensus among scholars is that scientific theories are inseparable from the context of their production and use (see Rouse, 1996). Science, according to these naturalistic studies, is an active process of making sense of the natural world that is dependent upon multiple conceptual and empirical elements—elements that are discipline and even theory specific and that make sense only when taken as a whole— a view that goes well beyond the narrow conceptions of science currently looked to in education.

This last statement, with its reference to the "active process," points out the dual nature of scientific knowledge that, we believe, has important pedagogical implications. What students encounter in the classroom is often presented as a kind of final-form knowledge—a "rhetoric of conclusions," as Schwab referred to it (1962). Even when the tentative nature of knowledge in science is acknowledged, the tendency has been to portray that knowledge as a static body of thought—the most recent final form—subject to change as new empirical evidence comes to light. Educators and researchers have given little attention to the instrumental role that scientific knowledge plays in inquiry and how that role might be conveyed to students.

Clearly scientific theories, as sets of models, contribute to human understanding of the natural world; and that understanding can only be complete if those theories are set in the broader context of the more or less implicit assumptions about the world and the methodologies they entail. This backward-looking view of science, however, is only part of the picture. Scientific theories, with their assumptions and especially their methodological prescriptions, are also, and perhaps more important for the research scientist, instruments used to meet the cognitive goals of the scientific community. These research programs (what might be thought of loosely as Kuhnian paradigms) structure the way one looks at and investigates what remains unexplained in nature; they circumscribe what phenomena can legitimately be made the subject of inquiry and suggest the path that the inquiry might take.

This heuristic value of scientific models has played an important part in the development of more adequate conceptions of scientific practice. For Kitcher, the advancement of science includes not only conceptual progress (more adequate specification of the referents of a model) and explanatory progress (greater unification of phenomena), but also what he termed "erotetic" progress, the ability of a model to suggest a range of interesting and tractable new questions not considered by previous models (1993). In this sense, a model always provides both an explanation for some specific phenomenon, and is itself the starting point for further elaboration, acting in this capacity as a tool for exploring the unknown.

Conclusion

How might these newer views of science help overcome the difficulties science education is confronting in evolutionary biology? To begin with, we have seen in the parallels between the historical reception of Darwin's theory and the conceptual problems of students evidence that the transition from pre-Darwinian ideas concerning the diversity of organic form to Darwin's theory of natural selection was not just a transition to a new theory, but rather a transition from one model of doing science to another. According to Kitcher, Darwin changed the "practice of biology" that included not only recasting what were seen as the primary problems in biology, but also "a methodological thesis about the adequacies of certain kinds of scientific theories" (1993, p. 34). This change entailed the adoption as well of a particular metaphysical world view that proved difficult for some to accept. Darwin's theory of natural selection therefore must be seen as an explanation embedded in a naturalistic world view that derived its logical warrant from its ability to unify and explain a wide range of phenomena within the discipline of biology - a method fundamentally different from that of physics. Darwin's theory cannot be understood properly apart from this context.

A more important difficulty student are likely to encounter in learning evolution, and even science more generally, can be anticipated from the accounts of science described above. Scientific theories, complex as they are with their multiple, interdependent elements, are constructed by the scientific community precisely to fill the dual role of explanation and exploration - to make sense of what is known and to guide future inquiry. Science instruction that ignores the latter function significantly handicaps student understanding. The introduction of Darwin's evolutionary theory not only explained the existing diversity of life on earth, but opened up a new set of questions concerning the living world in a variety of disciplines that had not been considered previously— questions that were both significant and tractable given the Darwinian model as a tool of inquiry. Conceptual understanding comes not from mere knowledge acquisition, but rather from the instrumental use of knowledge as a means to an end. Thus, the validity of Darwin's model cannot be passively demonstrated for students, but can only be fully realized in use. Such use, if it is to be effective in the context of scientific practice, will require students to become familiar with the metaphysical assumptions and methodological process that Darwin laid out. Theoretical context and scientific practice, in this view, are not just interdependent, but really two views of a single entity. It seems clear in this light that instruction or research that fails to address directly such contextual issues in evolution, regardless of other modifications, will continue to be inadequate.

The importance of evolutionary theory to the biological sciences is well established. Despite a recent push in science education research, the development of instruction that ensures student mastery of Darwin's work remains a daunting task. As we have attempted to argue here, we believe that much of the difficulty can be traced to a narrow conception of science rooted in the physics-based empiricism of the 19th century. The historical reception of the *Origin of Species* and the continued difficulty students have with the theory it describes provide compelling evidence for this claim. Problematic for both students and Darwin's contemporaries was and is the conflict between the prevailing metaphysics and that inherent in Darwin's view of organic change—a conflict compounded by an unorthodox methodology that remains poorly understood even today. Drawing on more recent models of science, those that include methodological, metaphysical, and social components as fundamental constituents of practice, allows one to see student alternative conceptions not as impediments to learning, but rather as important aspects of scientific practice to be embraced and engaged in a meaningful way. Perhaps through this broader conception of science, in which metaphysics and methodology can be legitimately and directly addressed, progress can be made in both research and instruction in evolutionary biology.

REFERENCES

Behe Michael, (1996). "Darwin's Black Box," Baltimore, MD: Johns Hopkins University Press.

Brumby, M.N. (1984). Misconceptions about the concept of natural selection by medical biology students. *Science Education, 68,* 493–503.

Coleman, W. (1977). *Biology in the nineteenth century: Problems of form, function, and transformation.* New York: Cambridge University Press.

Cummins, C.L., Demastes, S.S., & Hafner, M.S. (1994). Evolution: Biological education's under-researched unifying theme. *Journal of Research in Science Teaching, 31,* 445–448.

Darwin Charles (1859) "On the Origin of Species by Means of Natural Selection, or the Preservation of Favoured Races in the Struggle for Life," p. 162.

Darwin, C. (1873). *The origin of species by means of natural selection* (6th ed.). London: John Murray.

Demastes, S.S., Good, R.G., & Peebles, P. (1995). Students' conceptual ecologies and the process of conceptual change in evolution. *Science Education, 79,* 637–666.

Demastes, S.S., Good, R.G., & Peebles, P. (1996). Patterns of conceptual change in evolution. *Journal of Research in Science Teaching, 33,* 407–431.

Demastes, S.S., Trowbridge, J.E., & Cummins, C.L. (1992). Resource paper on evolution education research. In R.G. Good, J.E. Trowbridge, S.S. Demastes, J. H. Wandersee, M. S. Hafner, & C.L. Cummins (Eds.), *Proceedings of the 1992 Evolution Education research conference,* (pp. 42–71). Baton Rouge: Louisiana State University.

Michael Denton (1986), "Evolution: A Theory in Crisis," p. 250.

Dobzhansky, T. (1973). Nothing in biology makes sense except in the light of evolution. *American Biology Teacher, 35,* 125–129.

Donovan, A., Laudan, L., & Laudan, R. (Eds.). (1988). *Scrutinizing science: Empirical studies of scientific change.* Baltimore, MD: Johns Hopkins University Press.

Duschl, R.A., & Gitomer, D.H. (1991). Epistemological perspectives on conceptual change: Implications for educational practice. *Journal of Research in Science Teaching, 28,* 839–858.

Eve, R.A., & Dunn, D. (1990). Psychic powers, astrology, and creationism in the classroom? *American Biology Teacher, 52,* 10–21.

Galison, P., & Stump, D.J. (Eds.). (1996). *The disunity of science: Boundaries, contexts, and power*. Stanford, CA: Stanford University Press.

Giere, R.N. (1988). *Explaining science: A cognitive approach*. Chicago: University of Chicago Press.

Hewson, P.W., & Thorley, N.R. (1989). The conditions of conceptual change in the classroom. *International Journal of Science Education, 11*, 541–553.

Hodson, D. (1996). Laboratory work as scientific method: Three decades of confusion and distortion. *Journal of Curriculum Studies, 28*, 115–135.

Hull, D.L. (1973). *Darwin and his critics: The reception of Darwin's theory of evolution by the scientific community*. Cambridge, MA: Harvard University Press.

Illustra Media, (2002). Unlocking the Mystery of Life,"

Jenkins, E.W. (1996). The "nature of science" as a curriculum component. *Journal of Curriculum Studies, 28*, 137–150.

Joergensen, J. (1970). The development of logical empiricism. In O. Neurath, R. Carnap, & C. Morris (Eds.), *Foundations of the unity of science* (Vol. 2, pp. 847–935). Chicago: University of Chicago Press.

Kitcher, P. (1993). *The advancement of science: Science without legend. Objectivity without illusions*. New York: Oxford University Press.

Laudan, L. (1984). *Science and values: The aims of science and their role in scientific debate*. Berkeley, CA: University of California Press.

Mayr, E. (1991). *One long argument: Charles Darwin and the genesis of modern evolutionary thought*. Cambridge, MA: Harvard University Press.

Nyhart, L.K. (1995). *Biology takes form: Animal morphology and the German universities, 1800–1900*. Chicago: University of Chicago Press.

Pickering, A. (1995). *The mangle of practice: Time, agency, and science*. Chicago: University of Chicago Press.

Posner, G.J., Strike, K.A., Hewson, P.W., & Gertzog, W.A. (1982). Accommodation of a scientific conception: Toward a theory of conceptual change. *Science Education, 66*, 211–227.

Ruse, M. (1979). *The Darwinian revolution*. Chicago: University of Chicago Press.

Russell, E.S. (1916). *Form and function: A contribution to the history of animal morphology*. London: John Murray.

Schwab, J.J. (1962). The teaching of science as enquiry. In P.F. Brandwein (Ed.), *The teaching of science* (pp. 3–103). Cambridge, MA: Harvard University Press.

Siegel, H. (1993). Naturalized philosophy of science and natural science education. *Science and Education, 2,* 57–68.

Smith, M.U., Siegel, H., & McInerney, J.D. (1995). Foundational issues in evolution education. *Science and Education, 4,* 23–46.